The Ultimate "WHY" Book

Vol. 1 - The Sky

Parents Edition

Parents expert guide in explaining "WHY" for curious kids

LILO LU

Preface

Dear Parents, welcome to **The Ultimate "WHY" Book**!

As a parent, you've probably noticed that your child's curiosity knows no bounds. They ask endless questions, eager to understand the world around them! That curiosity is a powerful tool for learning, and it's something we should nurture and celebrate.

This book is designed to help you do just that. **The Ultimate "WHY" Book** is filled with the kinds of questions kids love to ask, like why the sky is blue, how rainbows form, or what makes the stars twinkle. Each answer is presented in a way that's engaging, fun, and easy for young minds to grasp.

But we know that sometimes, those simple answers spark even more questions! That's why we've created a special **Parent's Edition** of this book. This guide is meant to be read alongside your child, offering deeper insights and explanations to help you answer those follow-up "whys."

By exploring this book together, you're not just answering questions—you're fostering a love of learning that will last a lifetime. You're encouraging your child to be curious, to seek out new knowledge, and to find joy in understanding the world. So, let's dive in together and start this exciting journey of discovery.Who knows? You might find that you learn something new along the way too!

Contents

Why is the sky blue?

a) The Sky is blue because the ocean reflects its colour.

b) The Sky is blue because the blue fairy has spread its fairy dust all over the Sky.

c) The Sky is blue because of the way sunlight interacts with Earth's atmosphere.

Explanation:

The Sky's blue color is because of something called Rayleigh scattering, named after the British scientist Lord Rayleigh, who first explained it in the 19th century. It happens when sunlight mixes with the air around Earth.

1. What is Light?

Sunlight is like a rainbow in disguise – it has all the rainbow colors hidden inside. These colors have different lengths, like strings. The shorter strings are colors like blue and violet, while the longer ones are red and orange. When all these colors mix, we see white light, like the light we get from the sun.

2. What Happens in the Air?

When sunlight enters Earth's air, it bumps into tiny particles, like nitrogen and oxygen, which are the main ingredients in the air. The shorter color strings, like blue and violet, get scattered all around much more

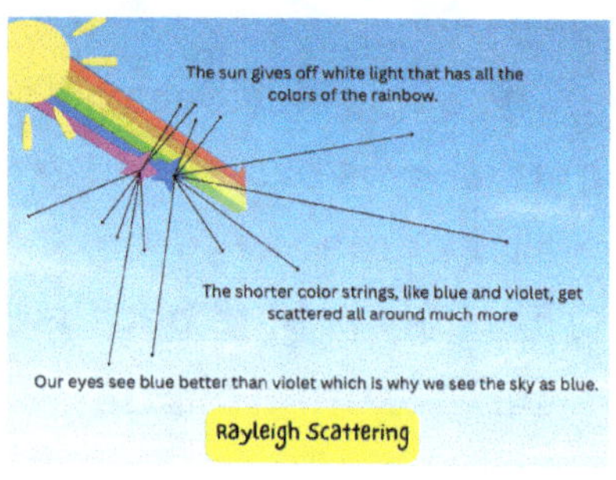

than the longer ones. This scattering makes the Sky colorful, as it spreads light across the Sky instead of letting it pass straight through.

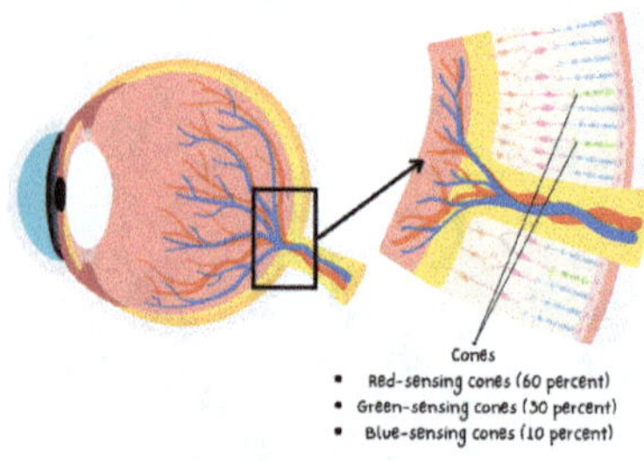

3. Why Do We See Blue?

Even though violet is scattered more than blue, we see the Sky as blue because our eyes are better at seeing blue. Our eyes have special sensors called cones, a type of photoreceptor cell in the retina that gives us our color vision. Cones are concentrated in the center of the retina in an area called the macula, which helps us see fine details.

The retina contains approximately 120 million rods and 6 million cones. There are three types of cone cells: red-sensing cones (60%), green-sensing cones (30%), and blue-sensing cones (10%). Since our cones are particularly sensitive to red, green, and blue light but not as much to violet, and because some violet light is absorbed by the air, what we mostly perceive is blue.

4. Why does the Sky change color?

The Sky doesn't stay the same color all day. The sun is lower in the Sky at sunrise and sunset, so its light has to travel through more air. This extra distance scatters away most of the blue light, leaving behind the red, orange, and yellow colors that make the Sky look like a beautiful painting during these times.

5. Proving the Science

Scientists have shown that shorter color strings (like blue) get scattered way more than longer ones (like red). This consistent observation explains why we see the Sky as blue during the day. The science behind Rayleigh scattering is well-established and has been confirmed through many experiments and observations.

This phenomenon not only explains the color of the Sky but also helps us understand other natural events, like why the ocean appears blue, why distant mountains look bluish, and why we see red sunsets. It's a key piece of how we understand light and its interaction with the world around us.

> *Another* **Interesting Fact:** Did you know that astronauts in space see a completely black sky, even when the sun is shining? This happens because there's no atmosphere in space to scatter the sunlight, so without the scattering effect, the Sky doesn't turn blue – it stays dark, even during the day!

Bonus Experiment: Make Your Own Blue Sky in a Jar

Objective: To help kids understand why the Sky looks blue by simulating the scattering of light using water, milk, and a flashlight.

What You'll Need:

- A clear glass jar or large, clear glass cup

- Water

- A few drops of milk or a small amount of powdered milk

- A flashlight (a phone flashlight works well)

- A dark room or a space where you can dim the lights

1. Fill the Jar: Filling the jar with water creates a medium that light can pass through, similar to how light from the sun passes through Earth's atmosphere. The water in the jar represents the atmosphere around us.
"What do you think will happen when we shine the light through the water? How does this compare to sunlight passing through the Sky?"

2. Add the Milk: Adding a few drops of milk to the water introduces tiny particles that will scatter the light, similar to how air molecules scatter sunlight. The milk makes the water slightly cloudy, which is important for creating the effect that simulates the blue Sky.
"Why do you think we're adding milk to the water? How do you think

it will change the way the light passes through the water? What does the cloudy water remind you of?"

3. Find a Dark Spot: Moving the jar to a dark room or dimming the lights allows you to see the effect more clearly. The darkness makes it easier to observe how the light interacts with the milk particles in the water.
"Why do you think it's important to turn off the lights? How do you think this will help us see what's happening inside the jar?"

4. Shine the Light: Shining the flashlight through the side of the jar allows the light to scatter as it passes through the cloudy water. The scattering of light, particularly the shorter blue wavelengths, makes the water appear bluish—similar to how the Sky looks blue during the day.
"What color do you see in the water when the light shines through it? How does this remind you of the color of the Sky?"

5. Observation and Explanation: The blue light scattering in the jar simulates how the Sky appears blue. The tiny particles of milk scatter the shorter blue wavelengths of light more than the longer red wavelengths, just like air molecules do with sunlight.
"Why do you think the water looks blue where the light passes through? How is this like what happens in the real Sky?"

6. Extra Fun (Optional): Shining the light at different angles through the jar can create effects similar to sunrise or sunset, where the light appears red or orange. This happens because the light has to pass through more of the cloudy water, similar to how sunlight travels through more of the atmosphere during sunrise or sunset.
"What colors do you see when we shine the light at an angle? How does this remind you of the colors you see in the Sky at sunrise or sunset?"

7. After the Experiment: Talk with the child about how the experiment simulates the scattering of light that makes the Sky appear blue. Explain that the tiny particles in the atmosphere scatter the shorter blue wavelengths of light more effectively, which is why we see a blue sky.

"What did you learn about why the Sky looks blue? How does this experiment help us understand what happens when sunlight passes through the atmosphere?"

8. Encourage Further Exploration: Suggest trying the experiment with different amounts of milk or at different times of day to see how it affects the color and brightness of the water. This can help the child understand how different conditions affect the appearance of the Sky. *"What do you think would happen if we added more or less milk? How might the color of the water change?"*

9. Reinforce Learning: Connect the experiment to real-world observations, such as why the Sky looks different at sunrise, noon, and sunset. Discuss how this understanding helps us appreciate the beauty of the natural world.

"Why do you think the Sky changes color throughout the day? How does this experiment help us understand those changes?"

Safety Note: Please remember to handle the flashlight and glass jar with care!

Why do clouds form?

a) Clouds form when warm air rises, cools down, and the water vapor in it condenses into tiny droplets.

b) Clouds form when the sky is stretching.

c) Clouds form when air pressure pushes water from the oceans into the sky.

Explanation:

Clouds are an important part of how water moves around Earth and how our climate works. Making a cloud starts with water turning into vapor and then into tiny water droplets, all influenced by air pressure and temperature.

1. Evaporation and Water Vapor

Clouds start forming when water from oceans, lakes, rivers, and even plants turns into an invisible gas called water vapor. This happens when the sun heats the water, causing it to rise into the air because it's lighter than the surrounding air. The warmer the air, the more water vapor it can hold, like a sponge that can soak up more water when it's warm.

2. Rising Air and Cooling

As the water vapor rises higher into the sky, it enters areas with lower air pressure and cooler temperatures. When the air rises, it spreads out and cools down, which is important for making clouds. Cooler air can't hold as much water vapor, so the vapor starts to change back into tiny water droplets in a process called condensation.

3. Condensation and Cloud Formation

Condensation happens when water vapor cools and turns back into liquid droplets. These droplets are super small, like tiny beads of water. For the water vapor to turn into droplets, it needs something to stick to, like dust or pollen in the air. These little particles are called **cloud condensation nuclei**. When enough droplets gather around the particles, they form a cloud.

4. Different Types of Clouds

The kind of cloud that forms depends on how high it is in the sky, the temperature, and how much water vapor is present. For example, fluffy white clouds, called **cumulus clouds,** usually form lower in the sky, where the air is warmer. **Stratus clouds**, which look like big gray blankets, form

in the middle of the sky. High up in the cold air, you'll find thin, wispy clouds called **cirrus clouds**, which are made of ice crystals instead of water droplets. Another type of cloud you might see is the **cumulonimbus cloud**. These are tall, towering clouds that can stretch from low altitudes to high in the sky. They often bring thunderstorms, heavy rain, and sometimes lightning. Cumulonimbus clouds are powerful and dramatic, and they can grow to great heights, even reaching the stratosphere!

5. Why Clouds Matter for Climate

Clouds are important for Earth's climate because they help control the temperature. During the day, clouds can reflect sunlight back into space, which cools the Earth like a big umbrella. At night, they act like a blanket, trapping heat and keeping the surface warmer. How clouds affect temperature depends on their type and height, making understanding clouds important for predicting climate changes.

Clouds are also influenced by big weather patterns around the world, like El Niño and La Niña, which can change the way air moves and, in turn, how clouds form across the globe.

Another **Interesting Fact:** Did you know that clouds can travel over 100 miles per hour? High-altitude clouds, like cirrus clouds, are often carried by powerful jet streams, allowing them to move incredibly fast across the sky, even though they may appear to drift slowly from the ground!

Bonus Experiment: Create a Cloud in a Jar

What You'll Need:

- A clear glass jar with a lid

- Hot water (use caution and assist with this step)

- Ice cubes

- A small plate or saucer

- Hairspray or aerosol air freshener (optional, with adult supervision)

1. Heat the Water: Pouring hot water into the jar creates warm, moist air inside, similar to how the Earth's surface heats up and causes water to evaporate, forming water vapor. The steam from the hot water represents the water vapor in the atmosphere.
"What do you think will happen when we add the hot water to the jar? Why do you think the water is steaming?"

2. Create a Cold Lid: Placing a small plate or saucer with ice cubes on top of the jar cools the air inside, simulating the cool upper atmosphere where clouds form. The cold lid causes the warm, moist air inside the jar to cool down and begin condensing into tiny droplets of water.
"Why do you think we're putting ice on top of the jar? How do you think the cold will affect the warm air inside?"

3. Wait and Watch: As the warm air rises and meets the cold lid, it cools down and begins to condense, forming tiny droplets of water. This process is similar to how clouds form in the sky. The condensation inside the jar creates a mist, which starts to look like a cloud.

"What do you see happening inside the jar? How does the mist forming inside the jar remind you of clouds in the sky?"

4. Add the Hairspray (Optional): Adding a small amount of hairspray or aerosol air freshener introduces tiny particles into the jar, which the water vapor can cling to, making the cloud more visible. In the atmosphere, dust and other particles serve a similar role, helping water vapor condense into droplets that form clouds.

"Why do you think we're adding hairspray? How do you think it helps make the cloud more visible?"

5. See the Cloud: After a few seconds, a visible cloud should form inside the jar. The cloud is made of tiny water droplets, just like real clouds in the sky. This step demonstrates how clouds are made up of countless tiny droplets of water.

"Can you see the cloud forming inside the jar? How does it compare to the clouds we see in the sky?"

6. Release the Cloud: Removing the plate allows the cloud to escape into the air, similar to how clouds move and change in the atmosphere. This step shows how clouds can disperse or change form over time.

"What happens to the cloud when we take off the plate? How does it change when it's released into the air?"

7. After the Experiment: Discuss the Science to explain how the hot water created warm, moist air that condensed into a cloud when it hits the cold lid, similar to how clouds form in the sky.

"What did you learn about how clouds are made? How does this experiment help us understand what happens in the sky?"

8. Encourage Exploration: Suggest trying the experiment again with different amounts of water, ice, or even using different sprays to see how it changes the cloud formation.
"What do you think would happen if we used more ice? How might that change the way the cloud forms? Do you want to try it?"

Safety Note: Make sure to do this experiment carefully especially when handling hot water.

Why is the cloud white?

a) Clouds are white because they reflect all the colors of sunlight equally.

b) Clouds are white because they are filled with snowflakes that reflect sunlight.

c) Clouds are white because they are made of tiny ice crystals that scatter blue light.

Explanation:

Clouds look white because light bounces off the tiny water droplets or ice crystals inside them. This process is known as Mie scattering and works differently from how the sky looks blue, which is due to Rayleigh scattering.

1. What Makes Light and Clouds

Sunlight, or white light, has all the colors of the rainbow mixed together. The light stays white when sunlight hits clouds, which are made up of tiny water droplets or ice crystals. These water droplets are just the right size to scatter all the colors of light equally, so when the light comes out of the cloud, it's still white.

2. Mie Scattering and Why Clouds Are White

Unlike the blue sky, which happens because shorter light waves (like blue) are scattered more, Mie scattering in clouds doesn't pick any favorite color. It scatters all colors

equally. So, when sunlight goes into a cloud, all the colors mix back together, and the cloud looks white to us.

3. Why Some Clouds Look Gray

The thickness of a cloud can change how bright or dark it looks. Thick clouds, like big storm clouds, might look gray because the light has to go through more water droplets or ice crystals. As the light travels through, some of it gets absorbed or scattered away, so less light comes out the bottom, making the cloud look darker. But most clouds, which aren't too thick, still look white because they let most of the light pass through.

4. How Sunlight Changes Cloud Colors

Clouds can change color depending on the sun's position and strength. At sunrise or sunset, clouds can turn red, orange, or pink because the shorter light waves (like blue and green) are scattered away, leaving behind the longer waves (like red and orange). But when the sun is high in the sky, clouds usually look white because the light is scattered evenly by the water droplets.

5. Why Cloud Whiteness Matters

Scientists study how clouds reflect sunlight, which is called cloud albedo. This reflection helps keep Earth cool by bouncing sunlight back into space. Whiter, more reflective clouds do better to cool the Earth because they send more sunlight away.

Another **Interesting Fact:** Did you know that the whiteness of clouds can be used to predict weather? Thicker, darker clouds often indicate impending rain or storms, while thinner, whiter clouds usually indicate fair weather. This is because the thickness of a cloud affects how much light it can scatter and transmit, giving us visual clues about the weather conditions.

Can I touch the sky if I go really high?

a) Yes, you can touch the sky if you go high enough in an airplane or a rocket.

b) No, you can't touch the sky because the sky is not a solid thing you can reach.

c) Maybe, but only if you climb the tallest mountain in the world.

Explanation:

The idea of touching the Sky sounds magical, but scientifically, the Sky isn't something you can touch. It's actually a visual effect created by the scattering of sunlight in Earth's atmosphere. The atmosphere is a layer of gases that surrounds our planet, and as you move higher, it becomes thinner until it eventually blends into space.

1. What is the Sky?

The "sky" we see is like a big, dome-shaped space above us, looking blue during the day and filled with stars at night. This happens because of something called Rayleigh scattering, where shorter wavelengths of light (like blue and violet) are scattered more than longer wavelengths (like red and orange). During the day, this scattering makes the Sky look blue, and at night, we see the dark space with stars.

2. The Layers of the Atmosphere

The atmosphere has several layers, each with different features:

- **Troposphere:** This is the closest layer to Earth's surface, going up about 8-15 kilometers (5-9 miles). It's where all the weather happens, clouds form, and planes fly.

- **Stratosphere:** Above the troposphere, this layer extends from 15 to 50 kilometers (9 to 31 miles). It has the ozone layer, which protects us by absorbing harmful UV radiation from the sun.

- **Mesosphere:** This layer goes from 50 to 85 kilometers (31 to 53 miles). It's where most meteors burn up when they enter the atmosphere.

- **Thermosphere:** Extending from 85 to 600 kilometers (53 to 373 miles), this is where the Northern and Southern Lights occur and where the International Space Station orbits.

- **Exosphere:** The outermost layer, starting around 600 kilometers (373 miles) and merging into space. There are very few air molecules here, and it gradually becomes outer space.

3. Can You Reach the Sky?

As you go higher through these layers, the air becomes thinner, and the blue Sky fades into the blackness of space. The Sky isn't a solid thing; it's just a perception made by the scattering of light. By the time you reach the thermosphere or exosphere, the air is so thin that the Sky, as we know it, disappears, and you enter space, where the Sky is always black, and stars are visible all the time.

4. What Scientists Say

Scientists understand that the Sky is an optical phenomenon, not something you can touch. The atmosphere, which creates the Sky, is vital for life on Earth. It protects us from harmful solar radiation and helps regulate temperature. But no matter how high you go, you won't find a spot where you can actually touch the Sky—it just transitions from the blue we see on the ground to the vast emptiness of space beyond our planet.

Another **Interesting Fact:** Did you know that Felix Baumgartner jumped from the stratosphere at an altitude of 39 kilometers (24 miles) above Earth in 2012? During his jump, he reached over 1,300 kilometers per hour (810 miles per hour) and became the first person to break the sound barrier without using a vehicle. Even from that height, he didn't "touch" the Sky but passed through different atmospheric layers before safely landing on Earth!

Why do rainbows appear?

a) Rainbows appear because the sun's rays are bent and split into different colors by raindrops.

b) Rainbows appear because clouds mix different colors in the sky after it rains.

c) Rainbows appear because the birds color the sky with crayons.

Explanation:

Rainbows are one of nature's most beautiful sights. They happen when sunlight interacts with water droplets in the air. The science behind rainbows involves three key steps: light bending, reflecting, and spreading out into different colors.

1. Light Bending (Refraction):

A rainbow starts forming when sunlight enters a water droplet in the air. As the light enters, it slows down and bends because it's moving from air (which is less dense) to water (which is more dense). This bending

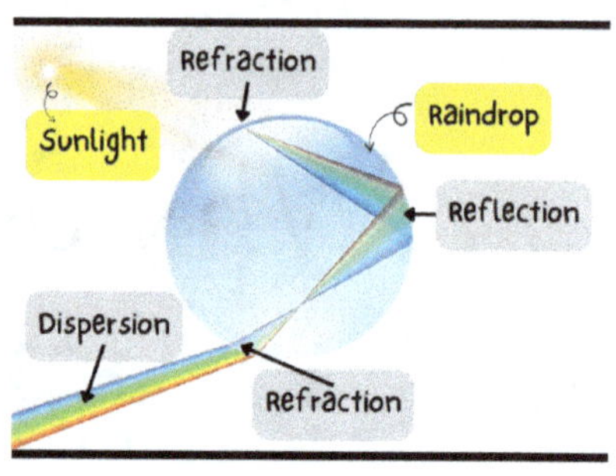

is called refraction. Sunlight is made up of many colors, and each color bends a little differently. Colors like blue and violet bend more than colors like red and orange, so the light starts to spread out into different colors.

2. Light Reflecting (Internal Reflection):

After bending, the light hits the back of the water droplet and bounces back inside. This bounce, called internal reflection, is significant because it sends the light back toward the front of the droplet. The light then bends again as it exits the droplet, which helps create the rainbow.

3. Spreading Out (Dispersion):

As the light exits the droplet, it bends again, and this further spreads out the different colors. This spreading out, called dispersion, creates the spectrum of colors we see in a rainbow, with red on the outer edge and violet on the inner edge.

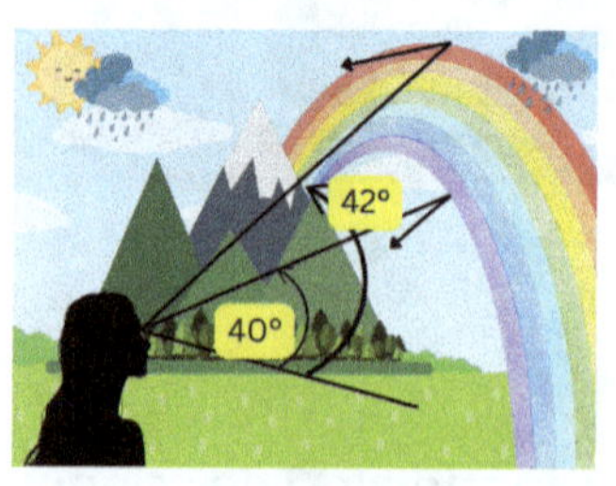

4. Seeing the Rainbow:

To see a rainbow, you need to stand with your back to the sun and look at a part of the sky where there's rain or mist. The light exits the droplets at a specific angle (usually around 42 degrees), giving the rainbow its curved shape. That's why rainbows are often seen in the morning or late afternoon when the sun is lower in the sky.

5. Double Rainbows:

Sometimes, the light bounces twice inside the water droplet before it exits. This creates a second, fainter rainbow above the first one, called a double rainbow. The colors in this second rainbow are reversed, with red on the inside and violet on the outside. It forms at a larger angle, around 50 to 53 degrees.

6. Why Rainbows Matter in Science:

Rainbows have helped scientists understand light and color. Isaac Newton's experiments with prisms showed that white light is made of different colors, leading to important discoveries in optics, the science of light. Studying rainbows also helps us learn more about how light behaves when it interacts with different materials.

Another **Interesting Fact:** Did you know that rainbows can form complete circles, but we usually only see a semicircle? The full-circle rainbow can sometimes be seen from high altitudes, like from an airplane or on a mountaintop. The ground typically blocks the lower part of the circle, which is why we usually see a rainbow as an arc.

Bonus Experiment: Create Your Own Rainbow

Objective: Allowing the child to better understand the science behind rainbow creation.

What You'll Need:

- A glass of water (clear glass)

- A small mirror

- A flashlight or access to sunlight

- A white piece of paper

1. Set Up the Experiment: Filling the glass with water and placing the mirror inside at an angle creates the basic setup needed to mimic the conditions that form a rainbow. The water acts as a medium for light refraction, while the mirror reflects the light, allowing it to be split into different colors.
"Why do you think we're placing the mirror at an angle? How do you think the light will change when it hits the water and the mirror?"

2. Find the Light Source: By positioning the glass near sunlight or using a flashlight, the light shines onto the mirror, which then reflects and refracts the light through the water. This setup is essential for splitting the light into its various colors, just like how sunlight creates a rainbow when it passes through raindrops.
"Why do you think the light needs to hit the mirror for a rainbow to appear? Why do you think we need to use sunlight or a flashlight?"

3. Catch the Rainbow: Holding the white piece of paper in the path of the light reflected off the mirror allows the child to see the rainbow that forms. The angle of the paper and the mirror determines how well the rainbow appears, mimicking the natural process of rainbow formation in the sky.

"What colors do you see in the rainbow? Why do you think the light changes into different colors? What happens when you change the angle of the paper or the mirror?"

4. Discuss the Science: Explain that the rainbow is created because the light bends (refracts) when it moves from air into water, and the mirror reflects the light, splitting it into different colors. This is similar to how rainbows form in the sky.

"What did you learn about how rainbows are made? Why do you think we see different colors in the rainbow? How does the light change when it passes through the water and hits the mirror?"

5. Encourage Exploration: Suggest trying different angles or varying the light source to see how the rainbow changes. This encourages further experimentation and understanding.

"What do you think would happen if we moved the mirror or changed the angle? Do you want to try it?"

6. Reinforce Learning: Connect the experiment to real-world rainbows by discussing when and where rainbows are most likely to be seen. Explain how raindrops in the sky act like tiny prisms, splitting sunlight into the colors we see in a rainbow.

"Why do you think we see rainbows after it rains? How do raindrops in the sky help create a rainbow like the one we made?"

Safety First!

Ensure that the child handles the glass and flashlight carefully. If using sunlight, make sure the child doesn't look directly into the sun.

Why can't I see rainbow at night?

<u>a) Rainbows only appear during the day because they need sunlight to form.</u>

b) Rainbows are shy and only come out when the moon is full.

c) Rainbows hide at night because they don't like the dark.

Moonbow and shining full moon. Credit: Wirestock/FreePik

Explanation:

Rainbows are beautiful and fascinating, but they only happen during the day. The reason is simple: rainbows need sunlight to form. Let's break down how this works.

1. How Rainbows Are Made

A rainbow appears when sunlight shines through water droplets in the air. As the light enters a droplet, it slows down and bends, a process called refraction. Different colors in the light bend differently because they have different wavelengths. This bending separates the colors, a process called dispersion. The light then bounces off the inside of the droplet (reflection) and bends again as it exits. This process happens in millions of droplets, creating a colorful arc in the sky that we see as a rainbow.

2. Why Sunlight is Important

Sunlight is the key ingredient for making a rainbow. The light needs to be strong and direct to go through all the bending and bouncing inside the water droplets. During the day, the sun provides this light. But the sun isn't there at night, so there's no sunlight to interact with the water droplets. Without sunlight, the rainbow-making process doesn't happen, which is why you can't see a rainbow at night.

3. Moonbows: Nighttime Rainbows

Even though traditional rainbows need sunlight, something called a "moonbow" or lunar rainbow can happen at night. Moonbows are made by moonlight instead of sunlight. But because the moon's light is much weaker than the sun's, moonbows are rare and usually very faint. They often appear white because the light is too dim for us to see the full range of colors. To see a moonbow, you need a nearly full moon, a dark sky, and rain or mist in the air opposite the moon.

4. Why is there no Rainbow at Night?

In short, you don't see rainbows at night mainly because there's no sunlight. The sun's rays are essential for creating the bending, separating, and reflecting of light that forms a rainbow. This process can't happen without the sun, so there's no rainbow. While moonbows can appear under special conditions, they're not the bright, colorful rainbows we see during the day but rather faint, ghostly arcs of light.

***Another* Interesting Fact**: Rainbows are not just an Earth-bound phenomenon. They have been observed on other planets and moons in our solar system. For example, rainbows can occur on Titan, one of Saturn's moons, where sunlight interacts with liquid methane droplets in the atmosphere, creating a "methane rainbow." This demonstrates that rainbows, while beautiful and familiar, are part of a broader set of natural processes that occur throughout the universe.

What happens if you find the end of a rainbow?

a) If you find the end of a rainbow, you'll discover a pot of gold hidden by leprechauns.

b) The end of a rainbow is just an illusion, so you can never actually reach it.

c) When you find the end of a rainbow, it disappears because rainbows are magical.

Explanation:

The idea of finding the end of a rainbow has long captured the imagination of people around the world. Tales of pots of gold guarded by leprechauns have made this concept a staple of folk-lore and fantasy. But what happens if you actually find the end of a rainbow? In reality, it's impossible to reach the end of a rainbow because rainbows don't have a fixed location. They are a visual effect created by the interaction between sunlight and raindrops, so their "end" is always a matter of perspective.

1. How Rainbows Form

To understand why you can't find the end of a rainbow, it helps to know how rainbows form. When sunlight enters a raindrop, it bends (refracts) and then reflects off the inside surface of the drop. As the light exits, it bends again, splitting into different colors because each color bends by a slightly different amount. This is called dispersion. The result is a spectrum of colors ranging from red (on the outer edge) to violet (on the inner edge).

Rainbows appear as circular arcs because of how light bends and reflects inside the water droplets. However, we typically see only part of the arc from the ground, so it looks like a half-circle stretching across the sky.

2. Rainbows and Perspective

The crucial thing to understand is that rainbows depend on your viewing angle. The sun needs to be behind you, and raindrops in the atmosphere need to be in front of you. The light that creates the rainbow is reflected at a specific angle—around 42 degrees from the direction opposite the sun. So, the rainbow you see is unique to your position and angle of view. Move a little, and the rainbow moves with you. This explains why the "end" of a rainbow always seems to be just out of reach. It's not a fixed point; it's a shifting optical illusion.

3. Why You Can't Reach the End

Rainbows are optical effects based on the observer's position; they don't have a physical location. If you try to move toward the rainbow, the angle between you, the sun, and the rain shifts, causing the rainbow to appear farther away or in a different position; in other words, the closer you try to get, the more the rainbow appears to move with you. It's like chasing a shadow—no matter how fast or far you run, the rainbow remains elusive because it's tied to your perspective.

4. Full Circle Rainbows

Sometimes, people in airplanes or at very high altitudes can see a full-circle rainbow. This happens because they can view the light refracted by raindrops in all directions, not just the half-circle visible from the ground. Even then, the rainbow is still an optical illusion with no tangible end.

Another **Interesting Fact:** Rainbows can sometimes appear as "double rainbows." This happens when the light bounces around twice inside the water droplets before coming out. The second rainbow is usually lighter and appears above the main rainbow. In a double rainbow, the second rainbow's colors are flipped, so the red is on the inside, and the violet is on the outside. The space between the two rainbows often looks darker, and this darker area is called "Alexander's band," named after a person who first noticed it a long time ago.

Why does the sun follow me?

a) The sun follows you because it's connected to your shadow.

b) The sun follows you because it likes you.

c) <u>The sun seems to follow you because it's so far away that it looks like it's in the same spot no matter where you go.</u>

Explanation:

Have you ever noticed that the sun seems to follow you wherever you go? Many kids notice this because of how far away the sun is and how we see things.

1. Distance and Perspective

The sun is really, really far away—about 93 million miles (150 million kilometers) from Earth! Because it's so far, the sun looks like it's in the same spot in the sky, no matter where you are. When you move just a little, like from one side of the playground to the other, that change is tiny compared to the huge distance to the sun. So, the sun seems to stay in the same place, making it look like it's following you. Even if you were to drive for miles or fly on an airplane, the sun would still seem to be in the same spot because of how far away it is.

2. Why Nearby Objects Look Different

When you move around, things that are close to you, like trees or buildings, seem to shift a lot compared to the background. This is called **motion parallax**. But because the sun is so far away, this effect doesn't happen with it. The sun's position hardly changes when you move, which is why it looks like it's staying right above you. In contrast, nearby objects seem to move more because they are much closer to you, making their movement more noticeable.

3. The Earth's Spin

The sun seems to move across the sky during the day, but this is actually because the Earth is spinning. As the Earth rotates, different parts of the world get sunlight. From where we stand, it looks like the sun is moving from east to west. But the sun isn't really moving around the Earth—it's the Earth spinning that creates this illusion, making it feel like the sun is always following you. This rotation is what causes the sun to "rise" in the morning and "set" in the evening.

4. How We See the Sky

The sky is huge, and the sun is a small, bright spot in it. Because the sun is so bright and far away, our brains see it as something that doesn't move much, even though we're moving. This makes it feel like the sun stays in the same place, following you wherever you go. Our brains are wired to recognize motion and distance in close-up objects, so when something as far away as the sun doesn't appear to change position, it can seem like it's keeping pace with us, no matter how far or fast we travel.

Another **Interesting Fact:** Did you know that the phenomenon of the sun appearing to follow you is similar to how the moon behaves? The moon, being much closer to Earth but still far enough away (about 238,855 miles or 384,400 kilometers), also seems to follow you as you move. This is because, like the sun, the moon's distance from Earth is so vast that its position in the sky doesn't change noticeably as you move around. This is why both the sun and the moon appear to be in a fixed position relative to you, giving the impression that they are following you!

Observation Activity: Why Does the Sun Follow Me?

Objective: To help kids understand why the sun appears to follow them when they're walking, driving, or moving around.

What You'll Need:

- A sunny day

- An open area outside where the sun is visible

- A small object, like a toy or a ball

- An adult to join the child in the activity

1. Go Outside: By going outside on a sunny day, you provide a clear view of the sun, allowing the child to observe how the sun appears in the sky. Ensuring it's a safe spot to observe without staring directly at the sun is crucial for protecting the child's eyes. *"Can you see the sun up in the sky? What do you think will happen if we start walking around—do you think the sun will stay in the same spot?"*

2. Start Moving: As the child walks around in a straight line or moves in a circle, they'll notice that the sun seems to stay in the same spot in the sky, regardless of their movement. This is because the sun is so far away that its position relative to it doesn't noticeably change.

"Does the sun look like it's moving, or does it seem to stay in the same place? Why do you think that is?"

3. Observe the Toy or Ball: Holding a toy or ball out in front while walking demonstrates the concept that objects close to you move with you while distant objects (like the sun) appear to stay in the same place. The toy represents the sun in this analogy, showing how proximity affects perception.

"Do you see how the toy stays with you as you move? Do you think the sun is doing something similar?"

4. After the activity: Explain that the sun is so far away that its position in the sky doesn't change much when we move around on Earth. This makes it look like the sun is following us, even though it's really just staying in the same spot in space.

"What did we discover about the sun today? Why does it seem like it's always in the same spot when we move?"

"Why do you think the sun looks like it's following us? What did we learn about how far away the sun is and how that affects what we see?"

6. Encourage Further Exploration: Suggest trying the activity at different times of the day, such as morning or evening, to see how the sun's position in the sky changes with time. This can help the child understand the daily movement of the Earth.

"Do you want to try this again later today to see if the sun looks different in the morning or evening? What do you think will happen?"

7. Reinforce Learning: Connect this activity to the concept of perspective in other areas, such as why mountains or buildings seem to follow us when driving or how distant stars and planets appear in the night sky.

"How does this idea of distance and perspective apply to other things

we see, like mountains or stars? Why do some things seem to move with us while others don't?"

Safety First!

Make sure the child understands not to look directly at the sun. Supervise their movements to ensure they stay in a safe area.

Why is the sun so hot?

a) The sun is so hot because it's made of burning coal like a giant campfire.

b) The sun is so hot because it's closer to Earth than any other star.

c) The sun is so hot because it's a big ball of gas where tiny particles smash together and release a lot of energy.

Explanation:

The sun's incredible heat comes from a process called nuclear fusion that happens deep inside its core. Here's how it works:

1. What the Sun is Made Of

The Sun is made up of different layers, each with a special job. Starting from the center, there's the **Core**, where the Sun's energy is created by turning hydrogen into helium. This process releases a lot of heat and light. The next layer is the **Radiative Zone**, where energy moves outward from the core by radiation. After that comes the **Convective Zone**, where energy is carried by the movement of hot gas, and you can see this as little bubbles on the Sun's surface.

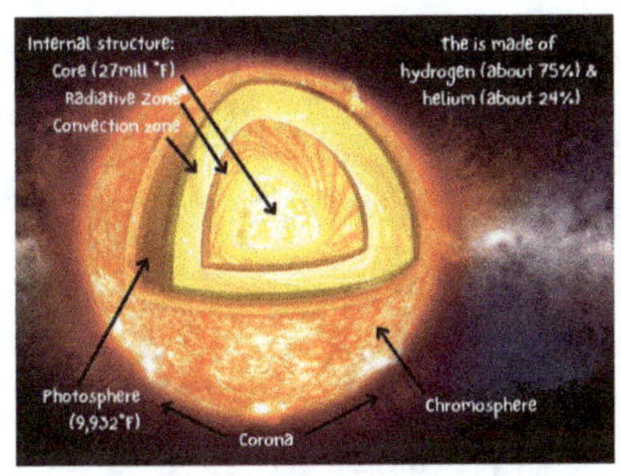

The Sun's atmosphere also has layers. The **Photosphere** is the surface we see when we look at the Sun. Above that is the **Chromosphere**, where the temperature gets much hotter. Then, there's the **Transition Region**, which separates the cooler parts from the **Corona**, the Sun's outer layer, which you can see during a solar eclipse.

The Sun is mostly made of two gases: hydrogen (about 75%) and helium (about 24%). At the very center of the Sun, the temperature is incredibly hot—around 15 million degrees Celsius (27 million degrees Fahrenheit). Here, the pressure is so strong that hydrogen atoms smash into each other, creating helium and releasing huge amounts of energy that we see as sunlight.

2. Nuclear Fusion: The Sun's Power Source

Nuclear fusion is like a super-powerful reaction where smaller atoms, like hydrogen, combine to form a

bigger atom, like helium. When this happens in the sun, a huge amount of energy is released. This energy comes from a tiny bit of the hydrogen's mass turning into energy, thanks to Einstein's famous equation, $E=mc^2$, which shows that even a small amount of mass can turn into a lot of energy.

3. How the Sun's Energy Travels

The energy created in the sun's core doesn't just stay there—it starts a long journey to the surface. First, it moves through the radiative zone, where it travels by radiation, like light passing through a window. Then, it goes through the convective zone, where hot gases bubble up like boiling water. Finally, the energy reaches the sun's surface, called the photosphere, and is sent out into space as light and heat, which we feel on Earth as sunlight.

4. The Sun's Surface Temperature

Even though the sun's surface, the photosphere, is cooler than the core, it's still incredibly hot—about 5,500 degrees Celsius (9,932 degrees Fahrenheit). The warmth we feel on Earth is the energy that started in the sun's core, travelled through its layers, and crossed the vast distance of space. This energy powers almost everything on Earth, from weather to plants growing through photosynthesis.

Another **Interesting Fact:** Did you know that the sun is so massive that it contains 99.86% of the total mass of the entire solar system? This immense mass gives the sun its powerful gravitational pull, which keeps the planets, moons, asteroids, and comets in their orbits around it. Despite being so far away (about 93 million miles or 150 million kilometers from Earth), the sun's gravity dominates the solar system, influencing the movement of all celestial bodies within it.

Unlock the Power of Sharing Knowledge

"The best way to find yourself is to lose yourself in the service of others." - Mahatma Gandhi

Did you know that sharing what you know can brighten someone else's world, just like when sunlight makes clouds look white? It's true! By sharing, we make things better for everyone, and here's your chance to do just that.

Would you help someone just like you—curious about how the world works, but unsure where to start?

My mission with *The Ultimate WHY Book* is to make understanding the big questions, like "Why is the sky blue?" or "Why do clouds form?", easy and fun for all kids (and maybe even grown-ups too!). I want every child to feel like a super scientist, discovering the wonders of the world around them.

But, to reach more curious minds, I need your help.

Most people pick books based on reviews, just like you might have done. That's why I'm asking you to help other kids and families who are eager to learn more about nature's coolest mysteries by leaving a review.

It doesn't cost anything and takes less than a minute, but your words could inspire someone to grab this book and go on their own journey of discovery. With your review, you could help...

...one more child get excited about science.
...one more parent find the perfect bedtime book.
...one more teacher share fun lessons in the classroom.
...one more kid explore experiments and ask "why" every day.

Want to make a difference? It's simple! Just scan the QR code below and leave your thoughts about *The Ultimate WHY Book for Curious Kids*. It can be just a few sentences, but it could spark a world of curiosity in others.

If you love helping others learn and explore, then you're exactly the kind of person I want to thank—so thank you, from the bottom of my heart!

With gratitude,
Lilo Lu

Why can't I look at the sun?

a) You can't look at the sun because it might make you feel too warm.

b) <u>You can't look at the sun because it's so bright that it can hurt your eyes.</u>

c) You can't look at the sun because it moves too fast across the sky.

Explanation:

Looking directly at the sun is very harmful because the sun gives off a lot of bright light and ultraviolet (UV) radiation that can seriously damage your eyes. The sun is much brighter than what our eyes can safely handle, and even a glance can cause lasting damage to your vision.

1. How Your Eyes Work

Your eyes are designed to control how much light comes in. The cornea and lens focus light onto the retina at the back of your eye, where special cells (rods and cones) turn the light into signals that your brain uses to create images. The iris ad-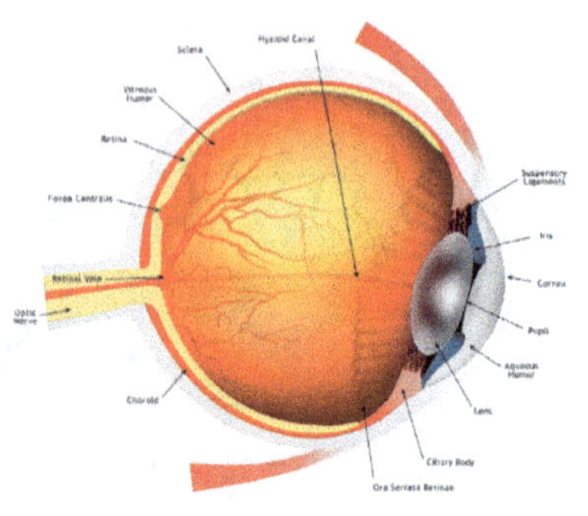justs the size of the pupil, letting in more light when it's dark and less light when it's bright. But when you look at the sun, it's so bright that your eyes can't adjust enough to protect you.

2. Damage from Light (Photochemical Damage)

When you look at the sun, too much light floods your eyes, and the UV rays can cause serious damage to the retina. This is called solar retinopathy. The UV light harms the cells in the retina that are crucial for seeing, which can lead to blind spots in your vision, either temporarily or permanently. Because the retina doesn't have pain sensors, you won't feel any pain when this damage happens, which makes it especially dangerous.

3. Heat Damage (Thermal Damage)

Besides the light damage, the sun's rays can also cause heat damage to your eyes. The focused light from the sun can actually burn the retina, destroying the tissue. This damage can happen very quickly, even in just a few seconds of looking at the sun, and once the retinal tissue is destroyed, it doesn't grow back, causing permanent vision loss.

4. The Role of UV Radiation

The sun sends out different kinds of light, including UVA, UVB and UVC rays. Although the Earth's atmosphere blocks some UV rays, a lot still reaches the ground. UVA rays can go deep into your eye and cause long-term damage, while UVB rays are mostly absorbed 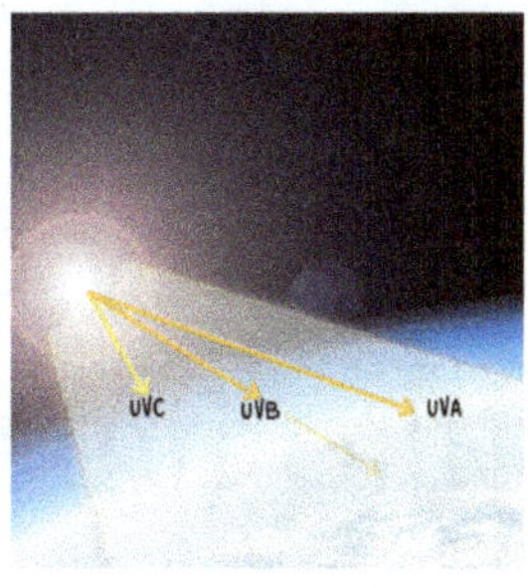

by the cornea and lens, leading to issues like cataracts and other eye problems.

5. Indirect Exposure

Even when you're not looking directly at the sun, like when sunlight reflects off water, snow, or shiny surfaces, it can still hurt your eyes. That's why it's important to wear sunglasses that block 100% of UV rays when you're outside, especially if you're around reflective surfaces.

Another **Interesting Fact:** Did you know that staring at laser pointers or bright lights without using proper eye protection can also caused solar retinopathy? There is no proven treatment for solar retinopathy, but many people get better on their own after 3-6 months.

Why does the sun look like it's being covered up?

a) The sun looks like it's being covered up because an eclipse happens when the moon turns into a giant cookie in the sky.

b) The sun looks like it's being covered up because the moon, Earth, or sun play a game of hide and seek, causing one to temporarily disappear.

c) The sun looks like it's being covered up because it's a magical event where stars decide to take a day off.

Explanation:

A solar eclipse is when it looks like the Sun is covered up, and this happens because the Moon moves directly between the Earth and the Sun, casting a shadow on Earth. Solar eclipses are exciting events that have amazed people for a long time. They happen because the Sun, Moon, and Earth line up just right.

1. What is a Solar Eclipse?

During a solar eclipse, the Sun can appear as a crescent, just like the Moon! A solar eclipse occurs when the Moon moves in front of the Sun, blocking some or all of its light from reaching Earth. There are three main types of solar eclipses:

- In a **total solar eclipse**, the Moon completely covers the Sun, making it dark, like nighttime, for a few minutes.

- In a **partial solar eclipse**, the Moon only covers part of the Sun.

- In an **annular eclipse**, the Moon covers the center of the Sun, leaving a bright ring around the edges, called a "ring of fire."

2. How Orbits and Distances Matter

A solar eclipse happens because of how the Moon orbits Earth and the distances involved. The Moon is about 384,400 kilometers (238,855 miles) away from Earth, and the Sun is about 150 million kilometers (93 million miles) away. Even though the Sun is about 400 times bigger than the Moon, it's also about 400 times farther away. This makes the Sun and Moon look almost the same size from Earth, allowing the Moon to cover the Sun completely during a total eclipse.

3. How Shadows Are Formed

When the Moon blocks the Sun's light, it creates two types of shadows on Earth:

- The **umbra** is the darkest part, where the Sun is completely blocked, causing a total eclipse.

- The **penumbra** is the lighter part, where only part of the Sun is covered, causing a partial eclipse.

The size and shape of these shadows depend on how the Sun, Moon, and Earth are lined up.

4. How Often Do Eclipses Happen?

Solar eclipses don't happen every month because the Moon's orbit is slightly tilted compared to Earth's orbit around the Sun. Most of the time, the Moon passes just above or below the Sun, so there's no eclipse. Solar eclipses happen about 2 to 5 times a year, but total eclipses are rare and can only be seen from a small part of Earth.

Another **Interesting Fact:** Did you know that the temperature can drop significantly within minutes during a total solar eclipse? The sudden lack of sunlight during a total eclipse can cause temperatures to drop by as much as 20 degrees Fahrenheit (11 degrees Celsius) or more, creating a brief but dramatic change in the environment around you. This temperature drop, combined with the darkening of the sky, makes a total solar eclipse an eerie and awe-inspiring experience.

Bonus Experiment: Create Your Own Eclipse at Home!

Objective: To help kids understand how a solar eclipse works by creating a simple model that simulates the positions of the Sun, Moon, and Earth during an eclipse.

What You'll Need:

- A flashlight (this will be the Sun)

- A small ball (like a ping-pong ball, this will be the Moon)

- A larger ball (like a basketball, this will be the Earth)

- A dark room

Steps:

1. Set Up the Sun: Turning on the flashlight and placing it on a table represents the Sun. The light from the flashlight simulates sunlight shining across space, just as the Sun shines across the solar system.

"What do you think will happen when the smaller ball (the Moon) passes in front of the larger ball (the Earth)? Do you think it will block the light?"

2. Position the Earth: Holding the larger ball about a foot away from the flashlight represents the Earth in its orbit around the

Sun. By positioning the Earth between the Sun and where you'll be standing, you're setting up the scenario for an eclipse to occur. *"Why do you think we need the Earth ball between the flashlight and the moon ball? How does this setup look like what happens in space?"*

3. Create the Eclipse: Moving the smaller ball (the Moon) in front of the larger ball (the Earth) simulates the Moon passing between the Earth and the Sun. As the Moon moves into position, it begins to block the light from the flashlight, casting a shadow on the Earth. This shadow represents the area of the Earth experiencing the eclipse. *"What do you notice happening to the light on the larger ball as we move the smaller ball in front of it?"*

4. Watch the Eclipse: Continuing to move the small ball until it completely covers the light from the flashlight creates a full eclipse. The Moon blocks the sunlight from reaching part of the Earth, just like during a real solar eclipse when the Moon's shadow falls on the Earth's surface. *"Can you see how the Moon (the smaller ball) blocks the Sun's light and creates a shadow on the Earth (the larger ball)? What do you think it would be like to be in that shadow on Earth during a real eclipse?"*

5. After the Experiment: Discuss the Science to explain how the experiment demonstrates the way a solar eclipse occurs when the Moon passes between the Earth and the Sun, blocking sunlight and casting a shadow on Earth. *"What did you learn about how eclipses happen? Why does the Moon's shadow cover the Earth during an eclipse?"*

6. Encourage Further Exploration: Suggest trying the experiment again with different distances between the balls and the flashlight or using different light sources to see how the shadow changes. This can help deepen the child's understanding of eclipses.

"What do you think would happen if we moved the moon ball farther away or closer to the Earth ball? Do you want to try it and see?"

7. Reinforce Learning: Connect the experiment to real-world eclipses by discussing when the next solar eclipse will occur and how to safely observe it. This can lead to a discussion about the importance of protecting your eyes during an eclipse.

"How do people safely watch a solar eclipse? Why is it important to protect our eyes when looking at the Sun?"

What's Happening?

In this experiment, the flashlight acts like the Sun, shining light across space. The larger ball is the Earth, and the smaller ball is the Moon. When the Moon moves between the Earth and the Sun, it blocks some of the sunlight, creating an eclipse. This is similar to what happens during a real solar eclipse in space!

Safety First: Ensure the child is handling the balls and flashlight safely and is comfortable with the setup. Keep the environment safe and distraction-free.

Why do shadows follow me everywhere?

a) Shadows follow you everywhere because they like to play with you.

b) Shadows follow you everywhere because the ground makes a picture of you as you move.

c) <u>Shadows follow you everywhere because when light shines on you, it makes a dark shape behind you.</u>

Explanation:

Shadows are something we see every day, but we might not think much about why they happen. A shadow is created when light hits an object and can't go through it, like when your body blocks sunlight. Here's how it works:

1. What Light Does

Light travels in straight lines from a source, like the sun or a lamp. When something solid, like your body, gets in the way, it blocks the light, creating a shadow. The shadow is just the dark area where the light can't reach because your body is in the way.

2. How Shadows Form

Shadows are made because light can't pass through solid objects. The size and shape of the shadow depend on the shape of the object and where the light is coming from. If the light is close to the object, the shadow looks bigger because the light spreads out more. If the light is far away, the shadow looks smaller and clearer because the light rays are more straight and parallel.

3. Why Shadows Follow You

Shadows seem to follow you because the light source, like the sun, stays in the same place, and your body keeps blocking the light in the same way as you move. So, as you walk, your shadow stays with you because your body keeps making the same block in the light. The shadow moves with you because you're always in the same position relative to the light.

4. How Shadow Length and Direction Change

Your shadow changes in length and direction depending on where the light is coming from. For example, when the sun is low in the sky, early in the morning

or late in the afternoon, your shadow is long because the light is coming at an angle. When the sun is high overhead at noon, your shadow is short. Even though the shadow's length and direction change, it always stays connected to you, making it seem like it's following you.

Another **Interesting Fact:** Did you know that astronauts on the moon experienced shadows differently from those on Earth? On the moon, shadows are much darker and sharper because there is no atmosphere to scatter the light. This effect, combined with the intense sunlight and the blackness of space, caused astronauts' shadows to appear pitch-black and sharply defined, creating a unique and somewhat eerie visual experience.

Why does the Moon change shape?

<u>**a) The Moon changes shape because of how we see the sunlight reflecting off it from Earth.**</u>

b) The Moon changes shape because it's made of cheese, and space mice nibble on it.

c) The Moon changes shape because it's a magical creature that can transform.

Explanation:

The Moon looks like it changes shape throughout the month because of its phases. These phases happen because of how the Moon orbits around Earth and the way sunlight hits the Moon's surface. A common mistake is thinking that Earth's shadow causes these phases, but that's not true. It's all about how sunlight shines on the Moon as it moves around Earth.

1. The Moon's Orbit

The Moon orbits Earth once every 27.3 days, but because Earth is also moving around the sun, it takes about 29.5 days for the Moon to go through all its phases. This time is called a lunar month. The Moon's path around Earth is slightly oval-shaped, but this doesn't change the phases we see much.

2. Phases of the Moon

The Moon doesn't make its own light; *it reflects sunlight.* The phases happen because of the changing angles between Earth, the Moon, and the sun. Half of the Moon is always lit by the sun, but from Earth, we see different parts of this lit-up side as the Moon moves.

- **New Moon:** During a new Moon, the Moon is between Earth and the sun, so the side facing us isn't lit up, and we can't see the Moon.

- **Waxing Crescent:** As the Moon moves, a small part of the lit side becomes visible, forming a crescent shape that slowly grows.

- **First Quarter:** When the Moon is at a right angle with Earth and the sun, we see half of the lit side, known as the first quarter.

- **Waxing Gibbous:** After the first quarter, more than half of the lit side is visible, leading to the waxing gibbous phase.

- **Full Moon:** The full Moon happens when the Moon is opposite the sun, so we see the entire lit side.

- **Waning Gibbous, Last Quarter, and Waning Crescent:** After the full Moon, the visible lit part starts to shrink (waning), moving through the waning gibbous, last quarter, and waning crescent phases until it becomes a new Moon again.

3. The Moon's Tilt

The Moon's orbit is tilted about 5 degrees compared to Earth's orbit around the sun. This tilt stops the Earth, Moon, and sun from lining up perfectly every month, which is why we don't see a lunar eclipse (when the Earth's shadow covers the Moon) every month.

Another **Interesting Fact:** Did you know that the same side of the Moon always faces Earth? This phenomenon, known as synchronous rotation or tidal locking, occurs because the Moon takes the same amount of time to rotate on its axis as it does to orbit Earth. As a result, we only see one hemisphere of the Moon from Earth, while the far side, often mistakenly called the "dark side," is never visible from our planet.

Can the Moon fall down to Earth?

a) Yes, the Moon could fall down to Earth if it gets too close.

b) No, the Moon can't fall down to Earth because it's held in place by gravity.

c) Maybe, but only if a big rocket pushes it towards Earth.

Explanation:

It might sound scary to think about the Moon falling to Earth, but it's very unlikely scientifically. The Moon stays in its orbit around Earth because of a careful balance between two essential forces: gravity and inertia. Let's break it down.

1. Gravity and How the Moon Moves

Gravity is the force that pulls things toward each other. Earth's gravity pulls on the Moon, constantly trying to bring it closer. If gravity was the only thing at work, the Moon would fall straight to Earth. But the Moon is also moving forward in its orbit, like a car going down a highway. This forward motion creates inertia, which is the tendency of an object to keep moving in a straight line unless something stops it.

2. Balancing the Forces

The Moon's inertia (its forward motion) works against the pull of Earth's gravity. Instead of falling, the Moon is in a kind of "freefall" around Earth, creating its orbit. Imagine swinging a ball on a string: the ball keeps moving in a circle because it's being pulled inward by the string (like gravity) while trying to move forward (like inertia). This balance keeps the Moon in a stable orbit around Earth.

3. Tidal Forces and the Moon's Orbit

Tidal forces, caused by the gravitational pull between Earth and the Moon, have a slow effect on the Moon's orbit. Right now, the Moon is moving slowly from Earth, about 1.5 inches (3.8 centimeters) per year. This happens because of a transfer of energy between Earth's rotation and the Moon's orbit. As Earth's rotation slows down a little due to tidal friction, the Moon gains some energy and moves slightly farther away.

4. The Moon's Future

Based on what we know about gravity and how orbits work, the Moon won't fall to Earth. Instead, it will keep slowly drifting away. This process will take billions of years, and long before anything significant happens, other cosmic events or changes in our solar system are likely to occur.

***Another* Interesting Fact:** Did you know that the Moon is responsible for creating the tides on Earth? The Moon's gravitational pull generates tidal forces, causing the oceans to bulge out in the direction of the Moon. This creates high tides in the areas facing the Moon and on the opposite side of the Earth, with low tides occurring in between. These tidal forces are also gradually slowing down Earth's rotation, which is why days are getting slightly longer over millions of years.

Can people live on the Moon?

a) Yes, people can live on the Moon because it has air and water like Earth.

b) No, people can't live on the Moon because there's no air to breathe or water to drink.

c) Maybe, but only if they bring their own food and build houses there.

Explanation:

The idea of living on the Moon has excited people for a long time, especially since the first Moon landing in 1969. While the Moon doesn't have any life, scientists are now more interested than ever in the possibility of humans living there. But there are big challenges we'd need to solve first.

1. The Moon's Tough Environment

The Moon's environment is very harsh and not friendly to humans. Unlike Earth, the Moon has no atmosphere to protect it from the sun's harmful rays, meteoroid impacts, and extreme temperatures. During the day, it can get as hot as 127 degrees Celsius (260 degrees Fahrenheit); at night, it can drop to minus 173 degrees Celsius (minus 280 degrees Fahrenheit). There's no air to breathe or protection from dangerous radiation, making it a risky place for humans.

2. Gravity and the Human Body

The Moon's gravity is only about one-sixth of Earth's. While jumping higher and moving more easily might sound fun, living in low gravity can be *harmful*. It can cause muscles to weaken, bones to lose density, and other health problems. Astronauts on the International Space Station face similar issues and need to exercise a lot to stay healthy.

3. Life Support Systems

To live on the Moon, people would need systems to provide air, water, food, and protection from radiation. Scientists are working on ways to get these resources on the Moon, like extracting water from ice and making oxygen from the soil. But these technologies are still being developed, and creating a safe, self-sustaining home on the Moon is a big challenge.

4. Mental and Social Challenges

Living on the Moon would also be tough on the mind. Being isolated, stuck in small spaces, and doing the same things every day could lead to mental health problems. The distance from Earth would make it hard to stay in touch with family and friends, and there would be long delays in communication. These issues would need to be carefully managed to keep people healthy and happy.

5. Current and Future Plans

Right now, no one lives on the Moon. However, space agencies like NASA have plans to send people there in the next few decades. NASA's Artemis program aims to land "the first woman and the next man" on the Moon by the mid-2020s and eventually build a base there, which could help with future missions to Mars.

Another **Interesting Fact:** Did you know that the Moon has water ice? Recent discoveries by NASA's Lunar Reconnaissance Orbiter and India's Chandrayaan-1 mission found evidence of water ice in permanently shadowed craters at the Moon's poles. This water could potentially be used for drinking, growing food, or even producing rocket fuel, making it a crucial resource for future lunar colonists.

How do stars get up in the sky at night?

a) Stars get up in the sky at night because they wake up when the sun goes to sleep.

b) Stars are placed in the sky every night by invisible fairies.

c) <u>Stars are always in the sky, but we can only see them at night when the sky is dark.</u>

Explanation:

The stars we see at night are huge, distant objects that have been shining for millions or billions of years. They aren't "placed" in the sky; we see them because of Earth's rotation and the fact that there's no sunlight at night.

1. What Are Stars?

Stars are giant balls of hot, glowing gas, mostly made of hydrogen and helium. They produce light and heat through a process called nuclear fusion, where hydrogen atoms combine to form helium, releasing a lot of energy. This is what makes stars shine. Our sun is just one of these stars, but there are billions more in our galaxy, the Milky Way, and even more in the entire universe.

2. Why We See Stars at Night

We see stars at night because of Earth's rotation. Earth spins on its axis once every 24 hours, giving us day and night. During the day, the side of Earth facing the sun is so bright that the light from stars is hidden. But as Earth rotates and the sun sets, the sky gets dark, and the stars become visible.

3. Why Stars Seem to Move

As Earth keeps rotating, the stars appear to move across the sky from east to west. This isn't because the stars are moving, but because Earth is spinning. The stars are so far away that even though they're moving fast, they seem to stay in the same place from our point of view on Earth.

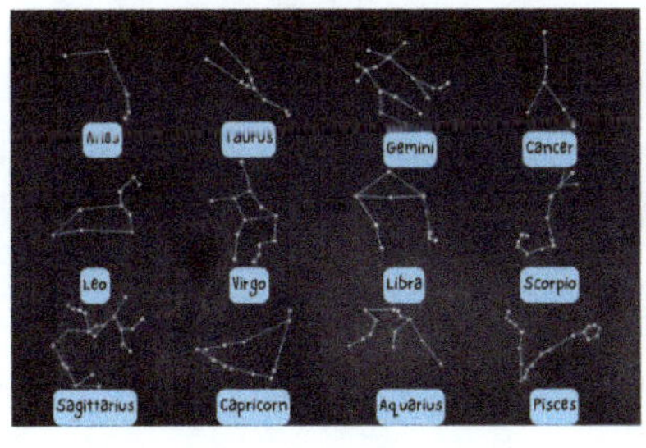

4. What Are Constellations?

People have noticed patterns in the stars and grouped them into constellations—shapes or figures that tell stories or help with navigation. These constellations seem to move across the sky during the night because of Earth's rotation, and they change positions with the seasons because of Earth's orbit around the sun.

5. Why We Don't See Stars During the Day

The stars are always shining, but the sun's light is so bright during the day that it hides the stars. When the sun sets, and its light fades, the stars become visible again.

***Another* Interesting Fact:** Did you know that some stars in the night sky are part of binary or even multiple star systems? This means that what appears to be a single star to the naked eye is actually two or more stars orbiting each other. In fact, about half of the stars you see in the sky are part of a system with two or more stars, which are gravitationally bound to each other and move together through space!

Why do stars twinkle?

a) Stars twinkle because their light passes through Earth's atmosphere, which makes it look like they are blinking.
b) Stars twinkle because they are moving around in space.
c) Stars twinkle because they are turning on and off like lights.

Explanation:

When you look up at the night sky, you might notice that stars seem to twinkle. This twinkling, called stellar scintillation, happens because Earth's atmosphere interferes with the light coming from the stars. Let's break it down.

1. How Starlight Travels

Stars send out light that travels across space for a long time before reaching us. This light stays steady as it moves through space. But because stars are so far away, they look like tiny points of light in the sky. When this light finally reaches Earth, it has to pass through our atmosphere, which can make it look like the stars are twinkling.

2. The Earth's Atmosphere

Our atmosphere is made of layers of air that have different temperatures, densities, and pressures. As starlight passes through these layers, it gets bent or refracted in different directions. Because the air is always moving, these layers constantly change, making the star's light bend in different ways from moment to moment. This bending causes the light to shift slightly, making the star appear to twinkle.

3. How Turbulence Affects Twinkling

Turbulence in the atmosphere, caused by wind and temperature changes, creates pockets of air that bend the starlight in unpredictable ways. The more turbulence there is, the more the star seems to twinkle.

4. Twinkling and the Star's Position

Stars closer to the horizon twinkle more than those directly overhead. This is because the light from stars near the horizon has to pass through more of the Earth's atmosphere, which means it encounters more turbulence and bending. Stars overhead have a shorter, more direct path through the atmosphere, so they twinkle less.

5. Why Planets Don't Twinkle

Unlike stars, planets don't usually twinkle when you look at them.

This is because planets are much closer to Earth and appear as small disks of light instead of tiny points. The larger size of planets makes them less affected by atmospheric turbulence, so their light stays steady.

***Another* Interesting Fact:** Did you know that astronomers use the absence of twinkling to identify planets? Since planets generally shine with a steady light, a celestial object that doesn't twinkle is likely a planet. This simple observation technique has been used for centuries to distinguish between stars and planets in the night sky!

Bonus Experiment: Why Do Stars Twinkle?

Objective: To help kids understand why stars twinkle by creating a simple model that mimics the effect.

What You'll Need:

- A flashlight or small LED light

- A clear glass of water

- A pen

- A piece of aluminum foil

- A dark room or a space where you can dim the lights

Steps:

1. Set Up the Star: By piercing several small holes in a piece of aluminum foil, you create tiny points of light that represent stars. Placing the foil in front of the flashlight allows the light to shine through these holes, simulating the appearance of stars in the night sky.
"Why do you think we're using small holes in the foil to represent stars? What do you think will happen when we shine the light through them?"

2. Prepare the Atmosphere: Adding a few drops of dish soap or milk to the water makes it slightly cloudy, which mimics Earth's atmosphere. The cloudy water represents the air and particles in our atmosphere that bend and scatter light, which causes the twinkling effect when we look at stars from Earth.

"What do you think will happen to the light when it passes through the cloudy water? How might this be similar to what happens to starlight?"

3. Observe the Twinkling: When you shine the flashlight through the foil and the cloudy water, the light gets bent and scattered by the water, just like starlight is bent and scattered by Earth's atmosphere. This scattering causes the light to flicker, mimicking the twinkling of stars in the night sky.

"Do you see how the light flickers as it passes through the water? How does this compare to the twinkling of stars you see in the sky?"

4. After the Experiment: Explain to the kids that in space, stars don't actually twinkle. They shine steadily. It's the Earth's atmosphere that makes them look like they're twinkling. When the air moves around, it bends the light from the stars in different ways, making them seem to twinkle.

"What did you learn about why stars twinkle? How does the Earth's atmosphere affect the light from stars?"

5. Encourage Further Exploration: Suggest trying the experiment again with different amounts of dish soap or milk in the water to see how it affects the twinkling effect. This can help the child understand how variations in the atmosphere might change the way stars appear.

"What do you think would happen if we made the water more or less cloudy? Do you want to try it and see how it changes the twinkling?"

6. Reinforce Learning: Connect the experiment to real-world astronomy by discussing how telescopes in space, like the Hubble Space Telescope, can see stars more clearly because they are above Earth's atmosphere. This can lead to a discussion about the importance of space exploration.

"Why do you think telescopes in space can see stars more clearly than those on Earth? How does being above the atmosphere make a difference?"

What's Happening?

This experiment shows why stars appear to twinkle. The cloudy water represents the Earth's atmosphere, which is full of moving air and tiny particles. As the light from the "star" passes through the atmosphere, it gets bent and scattered in different directions. This bending and scattering make the star's light appear to flicker or twinkle when we see it from Earth.

Safety First: Ensure the child handles the flashlight and water safely and supervise closely during the experiment, especially when in a darkened room.

Why does meteor shower happen?

a) Meteor showers happen when Earth passes through a trail of dust and rocks left by a comet.

b) Meteor showers happen because the Moon pulls pieces of the sky down with its gravity.

c) Meteor showers happen because the Moon is having a night-time disco party, and the stars are dancing too hard, throwing glitter everywhere!

Explanation:

Imagine Earth is like a car driving down a highway, and along this highway, there's a lot of space debris left behind by a truck (which represents a comet). This debris is like tiny pebbles, dust, and ice chunks that have fallen off the truck as it sped down the road. Now, when our car (Earth) drives through this debris field, the pebbles and dust hit the windshield (Earth's atmosphere) at incredibly high speeds. This is what we see as a meteor shower.

1. The Source of Meteors

The "truck" in our story is a comet. Comets are big, icy objects that travel around the Sun. When a comet gets close to the Sun, the heat causes it to melt, and pieces of it break off. These pieces spread out along the comet's path, creating a debris trail. This trail is like a breadcrumb path left by the comet, and Earth crosses it at the same time every year, causing a meteor shower.

2. What Happens When Earth Hits the Debris

When the tiny bits of debris, called meteoroids, hit Earth's atmosphere, they move so fast that they create a lot of friction with the air. This friction makes them heat up and glow, turning into bright streaks of light we call meteors or "shooting stars." But they aren't really stars—they're just tiny rocks burning up. Most of these meteoroids are so small that they burn up completely and never reach the ground. Sometimes, if a piece is big enough, it can survive the journey through the atmosphere and land on Earth. When that happens, it's called a meteorite.

3. Why Meteor Showers Happen at the Same Time Every Year

Earth's orbit is like a giant racetrack that circles around the Sun. Since Earth follows the same track every year, it crosses through the same debris fields left by comets at the same times each year.

That's why certain meteor showers, like the Perseids in August, happen every year like clockwork.

4. What Science Has Learned

Scientists have learned a lot about the stuff that makes up comets by studying meteor showers. When meteors burn up, they emit light, and by analyzing this light, scientists can figure out what elements are in the meteoroids. This helps them understand the materials that were around when our solar system was formed. Meteor showers also give scientists a way to track comets and make sure none of the debris paths pose a danger to Earth.

5. A Little Bit of Space Dust

When meteoroids burn up in our atmosphere, they release tiny amounts of metals like iron and magnesium. These tiny particles stay in the upper atmosphere for a long time, where they can affect things like radio signals, though the impact is usually minimal.

Another **Interesting Fact:** Sometimes, after a really bright meteor, you can see a "persistent train," like a glowing trail left in the sky. It's made of ionized gas and can last for several minutes, swirling around as high-altitude winds move it. This happens a lot during the Leonid meteor shower, where you might see these glowing trails even after the meteor itself is gone.

Why does it rain?

a) It rains because the sky is crying.

b) It rains because water vapor in the air cools down, condenses into droplets, and falls to the ground.

c) It rains because the wind blows water from the oceans into the clouds.

Explanation:

1. How Rain Forms

Think about when you boil water in a pot. You see steam rising from the water. This steam is water vapor, which is water in its gas form. In nature, the sun heats up water in oceans, lakes, and rivers, turning it into water vapor. This process is called evaporation. Every day, about 1 trillion tons of water evaporates from the Earth's oceans! The water vapor rises into the sky and mixes with the air, but it seems invisible since we can't see water vapor.

As the water vapor rises higher and higher, it enters cooler parts of the atmosphere. As the air gets cooler, the water vapor cools down too. When the water vapor cools, it turns back into tiny water droplets, just like when steam from a hot shower forms droplets on a cold mirror. This process is called **condensation**. These droplets group together to form clouds, which are made up of millions of tiny water droplets!

Now, just like when a sponge gets too full and starts dripping, the clouds can only hold so much water. When many water droplets join together, they get heavier and heavier. For it to rain, a droplet must grow about **a million times larger** than a single water molecule. Once the droplets become too heavy to stay in the cloud, they fall to the ground as rain. This is called **precipitation**.

Rain is part of the **water cycle**, which is nature's way of moving water from the ground to the sky and back again. Rain helps plants grow and fills up lakes and rivers and keeps everything alive by providing fresh water for drinking and irrigation. Without the water cycle, life on Earth wouldn't be possible!

2. How Scientists Understand Rain

Scientists use advanced technology like NEXRAD (Next-Generation Radar) to study and predict rainfall. NEXRAD works like a giant radar system that sends signals that bounce off raindrops, allowing scientists to detect where rain is falling and how heavy it is. Satellites orbiting high above Earth also monitor clouds, sending pictures and data back to scientists, helping them understand weather patterns and track how rain moves across the planet.

Rain is essential for keeping our environment healthy. It waters plants, fills rivers and lakes, and supports all kinds of life. But rain doesn't fall evenly everywhere. Some places get lots of rain all year round, while others have long dry periods with only occasional rain.

Another **Interesting Fact**: Did you know that some places on Earth can receive several meters of rain in a year, like the rainforests of the Amazon, while others, like deserts, get almost no rain at all? The wettest place on Earth is Mawsynram, India, where they receive over 11 meters of rain each year!

Why does the sky make thunder and lightning?

a) The sky makes thunder and lightning because stars are falling from space.

b) The sky produces thunder and lightning when it is upset about something.

c) The sky makes thunder and lightning when ice and water inside clouds collide, creating electric charges.

Explanation:

Imagine a big storm cloud as a giant battery in the sky. This battery has a lot of energy, with the top of the cloud being positively charged (like the positive end of a battery) and the bottom being negatively charged (like the negative end). This happens because, inside the cloud, tiny ice particles and water droplets bump into each other, causing some of them to lose electrons, which are tiny negatively charged particles. This creates a big difference in charge between the top and bottom of the cloud.

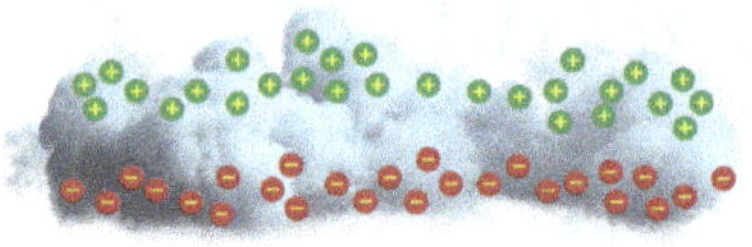

1. How Lightning Happens

When the difference in charge between the top and bottom of the cloud becomes too great, it's like the battery needs to be "recharged." The cloud releases this energy in the form of lightning, which is a huge spark of electricity. Lightning can strike within the cloud, between clouds, or from the cloud to the ground.

Think of lightning as a giant spark that tries to balance out the electrical charges. The lightning takes the quickest and easiest path it can find through the air, which is why it often looks zigzagged. When lightning happens, it heats up the air around it to an extremely high temperature, much hotter than the surface of the Sun. This sudden heating causes the air to expand rapidly.

2. How Thunder Is Made

Thunder is the sound that happens when lightning heats up the air. The heat causes the air to expand so fast that it creates a shockwave, similar to a balloon popping. This shockwave is what we hear as thunder. You usually see lightning before you hear thunder because light travels much faster than sound. It's like seeing fireworks in the sky before you hear the boom.

3. Why Thunder and Lightning Are Important

Scientists study thunderstorms to learn more about how lightning forms. Some researchers think that tiny particles from space, called cosmic rays, might help start lightning by making it easier for electricity to travel through the air. Understanding lightning better also helps scientists develop ways to protect people and buildings from lightning strikes, like using lightning rods.

4. Staying Safe During Thunderstorms

Thunderstorms can be dangerous, so it's important to stay safe. If you hear thunder, it means lightning is close enough to strike, so you should find shelter indoors. It's also a good idea to stay away from windows and avoid using electrical appliances until the storm passes.

Another **Interesting Fact:** Did you know that lightning doesn't always strike from the sky down to the ground? Sometimes, it actually goes from the ground up to the clouds! This happens when something tall, like a skyscraper or a mountain, has a strong enough electric charge to reach up and connect with the cloud. This type of lightning is called "ground-to-cloud" lightning, and it looks like the Earth is sending a bolt of lightning back up into the sky.

Science Observation Experiment: Thunder and Lightning

Objective: To help kids understand the relationship between thunder and lightning by observing and recording their observations during a storm.

What You'll Need:

- A notebook or paper

- A pencil or pen

- A clock or stopwatch

- A safe place indoors with a window view of the sky (like your living room or a covered porch)

Steps:

1. **Get Ready:** Being prepared and in a safe location ensures that you can observe the storm without any risk, and having your materials ready means you won't miss any important details.
   "Why is it important to stay indoors during a thunderstorm? What should we do if the storm seems to be getting closer?"

2. **Listen and Watch:** Lightning and thunder happen at the same time, but because light travels faster than sound, we see the lightning before we hear the thunder. Counting the seconds between the two helps estimate the distance of the storm.
"How long does it take for the sound of thunder to reach us after

seeing lightning? What do you notice about the time between the lightning and thunder as the storm continues?"

3. Record Your Observations: Recording the time between lightning and thunder allows you to analyze the data later to understand patterns and how the storm moves.

"Why is it important to write down our observations immediately after each lightning and thunder? How can our observations help us understand the storm better?"

4. Repeat: Repeating the observation gives you more information, which can confirm your findings and show how the storm is changing over time.

"Does the time between the lightning and thunder change as the storm moves?"

5. After the Observation: Review the observations with the kids and discuss what they learned about the relationship between lightning and thunder. Help the kids calculate how far away the storm was based on their observations. Explain that every 5 seconds between lightning and thunder means the storm is about 1 mile away.

"Did you notice any patterns in the timing between the lightning and thunder as the storm went on?"

6. Encourage Further Exploration: Ask if they noticed anything else about the storm and suggest other related experiments, like tracking how long the storm lasted or observing how different types of clouds relate to storms.

"What happens if you observe a different storm?"

What's Happening?

Lightning and thunder occur at the same time, but because light travels faster than sound, we see the lightning first. By counting the seconds between the lightning and the thunder, we can estimate how far away the storm is. Every 5 seconds between the flash and the thunder means the storm is about 1 mile (or 1.6 kilometers) away.

Safety note: Make sure that everyone is safely indoors, away from windows and glass doors. Remind the children about the importance of staying inside during a thunderstorm.

A BIG Thank You!

We hope you and your little ones have been loving **The Ultimate WHY Book**! We're all about answering those big "WHY" questions that kids seem to ask a million times a day, and we know that curiosity is a gift. But guess what? By leaving a review, you're not just helping us – you're helping so many other curious kids and their grown-ups discover the wonders of nature too!

So here's a question for you:

Have you ever found something amazing – like a great recipe or a super fun game – and felt that warm fuzzy feeling when you share it with a friend? That's exactly what you're doing when you leave a review. You're giving other families the chance to dive into a book that might just spark a lifelong love of learning and exploration in their kids.

Leaving a review for *The Ultimate WHY Book* is like putting a little extra magic into the world.

It's an opportunity to share your thoughts and experiences with other parents, caregivers, and kids who are just waiting to find their next favorite book. Imagine a child somewhere out there who's been asking "Why is the sky blue?" and is just itching for a fun, easy-to-understand explanation. Your review might be the nudge their parent needs to pick up this book and say, "Aha! Here's the answer!"

Now, we know your time is valuable (you've probably got a lot of WHY questions on your hands right now!), but by leaving a quick review, you'll be helping other families find a book that's packed with playful analogies, adorable pictures, and fun mini-experiments that make learning science super exciting. Your honest feedback will also help us improve, so we can keep creating books that inspire the next generation of scientists, thinkers, and dreamers.

How do you leave a review? We've made it super simple! Just scan the QR code below, and it'll take you directly to our review page. All you need to do is share your thoughts – it can be as simple or as detailed as you'd like. A few sentences can go a long way in helping others discover the fun inside *The Ultimate WHY Book*.

Your review makes a difference. When parents see real experiences from people like you, it helps them make decisions about what books to share with their kids. Reviews help us reach more families who are looking for a fun way to answer life's big questions, and they can inspire kids all over the world to stay curious about the world around them.

As a thank you, think about the joy you'll be spreading – not just to other parents, but to the kiddos who will soon be saying, "Wow! That's why the sky is blue!" You're introducing them to a world of discovery, and that's a gift that will keep on giving. Plus, in the world of writing and publishing, your support helps us grow and keeps us motivated to continue creating more awesome content for curious minds everywhere.

So, what do you say? Will you take a moment to leave a review and share the magic? We're excited to hear your thoughts, and we know others will be too. Just scan the QR code above to get started.

Thank you so much for being part of our curious community!

Happy learning and exploring!

References

Admin. (2022, May 31). Why doesn't the moon fall into the Earth? - CBSE Class 6-10. BYJUS. https://byjus.com/physics/why-doesnt -the-moon-fall-into-the-earth/

American Academy of Ophthalmology. (n.d.). What are cones? American Academy of Ophthalmology. https://www.aao.org/e ye-health/anatomy/cones

Ask an astronomer. (n.d.). Cool Cosmos. https://coolcosmos.ipac .caltech.edu/ask/210-Why-do-the-stars-twinkle

Big ideas: Why the sky is blue. (n.d.). NASA Science. https://scien ce.nasa.gov/learn/heat/big-ideas/big-idea-3-2/

Clouds and precipitation weather. (n.d.). How do clouds form? Clouds are made up of tiny water droplets or ice crystals. Slide-Player. http://slideplayer.com/slide/5981283/

Danielle. (2023, April 18). Make a cloud form in a jar science experiment. Cool Science Experiments Headquarters. https://coolsc ienceexperimentshq.com/make-a-cloud-form-in-a-jar/

Diagram of the eye. (n.d.). Harvard Eye Associates. https://harva rdeye.com/uncategorized/diagram-of-the-eye/

Earth's atmosphere: A multi-layered cake. (n.d.-b). NASA Science. https://science.nasa.gov/earth/earth-atmosphere/earth s-atmosphere-a-multi-layered-cake/

Education, U. C. F. S. (n.d.). The colors of the sky | Center for Science Education. UCAR. https://scied.ucar.edu/kids/sky-won ders/colors-sky

Facts about the moon. (n.d.). Actforlibraries.org. http://www.actf orlibraries.org/facts-about-the-moon-10/

Hiskey, D. (2013, June 21). Why the sky is blue. Today I Found Out. https://www.todayifoundout.com/index.php/2010/05/wh y-the-sky-is-blue/

How to explain the water cycle. (n.d.). Star Language Blog. https://www.starlanguageblog.com/how-to-explain-the-water-cycle/

How are rainbows formed? (n.d.). Met Office. https://www.metoffice.gov.uk/weather/learn-about/weat her/optical-effects/rainbows/how-are-rainbows-formed

Immelman, M. (2023, September 3). 5 easy science experiments for kids. KidsGearGuide. https://www.kidsgearguide.com/5-ea sy-science-experiments-for-kids/

JPL NASA. (n.d.). Sun and moon shapes. https://www.jpl.nasa.gov /edu/news/tag/Eclipse

Living on the moon - How to make it possible. (n.d.). HeroX. https://www.herox.com/blog/957-3-problems-we-nee d-to-solve-before-we-can-live-on

Marcy, A. (2024, April 12). Make sky in a jar! - Double Helix. Double Helix. https://blog.doublehelix.csiro.au/make-sky-in-a-jar/

Mie scattering. (n.d.). Apollo NVU. https://apollo.nvu.vsc.edu/clas ses/met130/notes/chapter19/mie_scatt.html

Moonbow shining full moon clear gray sky. (n.d.). Freepik. https://www.freepik.com/premium-photo/moonbow -shining-full-moon-clear-gray-sky_155191950.htm

NASA. (n.d.). The water cycle. NASA Global Precipitation Measurement. https://gpm.nasa.gov/education/water-cycle

Ochoa, C. (2023, January 11). How to make a rainbow with an experiment and craft. Little Passports. https://www.littlepassp orts.com/blog/craft-diy/how-to-make-a-rainbow/

Parallax. (n.d.). Wikipedia. https://en.wikipedia.org/wiki/Parallax

Popardowski, E., Hebda, T., & Jakubowski, T. (2022). The effect of UV-C irradiation on the mechanical and physiological properties of potato tuber and different products. Applied Sciences, 12(12), 5907. https://doi.org/10.3390/app12125907

Quora. (2017, June 15). Why can't we see stars during the day? Forbes. https://www.forbes.com/sites/quora/2017/06/15/why-cant-we-see-stars-during-the-day/

Rainbow. (n.d.). Ilmiosito. https://www.isabellafabbri.com/rainbow

Refraction of light. (n.d.). Science Learning Hub. https://www.sciencelearn.org.nz/resources/49-refraction-of-light

Saunders, T. (n.d.). How hot is the Sun? Our star's staggering temperature, explained. BBC Science Focus Magazine. https://www.sciencefocus.com/space/how-hot-is-the-sun/

Shooting star craft. (n.d.). Twinkl. https://www.twinkl.com.au/resource/paper-plate-shooting-star-space-crafts-t-tc-1664305795

Solar retinopathy. (n.d.). AAPOS. https://aapos.org/glossary/solar-retinopathy

Stars. (n.d.). NASA Science. https://science.nasa.gov/universe/stars/

Tales of rainbows and pots of gold. (n.d.). ConnollyCove. https://www.connollycove.com/tales-of-rainbows-and-pots-of-gold/

The color of clouds. (n.d.). National Oceanic and Atmospheric Administration. https://www.noaa.gov/jetstream/clouds/color-of-clouds

The science behind why the sky appears blue: Explained. (n.d.). Factober. https://factober.com/article/the-science-behind-why-the-sky-appears-blue-explained/

Theick. (2023, August 31). Why is the sky blue? LearnThought. https://learnthought.com/science/why-is-the-sky-blue/

The water cycle steps – How the water cycle works step by step. (n.d.). SmartClass4Kids. https://smartclass4kids.com/the-water-cycle/

Thoopal, R. K. (2005, August 18). Why does the sun follow you? Pitara Kids Network. https://www.pitara.com/science-for-kids/5ws-and-h/why-does-the-sun-follow-you/

Toppr. (2022, September 5). During scattering of light—the amount of scattering is inversely proportional to the wavelength of light—square, fourth power, cube, or half? Toppr Ask. https://www.toppr.com/ask/en-au/question/during-scattering-of-light-the-amount-of-scattering-is-inversely/

Umbra, penumbra, and antumbra. (n.d.). Wikipedia. https://en.wikipedia.org/wiki/Umbra,_penumbra_and_antumbra

Water cycle - Water | Term 3 Unit 2 | 6th Science. (n.d.). BrainKart. https://www.brainkart.com/article/Water-cycle_43134/

What is a solar eclipse? | NASA Space Place. (n.d.). NASA Space Place for Kids. https://spaceplace.nasa.gov/eclipse-snap/en/

Why can't you look directly at the sun? (2018, May 16). Wonderopolis. https://www.wonderopolis.org/wonder/why-cant-you-look-directly-at-the-sun

Why do stars twinkle? (n.d.). Nanogirl. https://www.nanogirl.co/science-experiments-videos/why-do-stars-twinkle

Why is the sky blue? (n.d.). NASA Space Place. https://spaceplace.nasa.gov/blue-sky/en/